YOUR KNOWLEDGE HAS VALUE

Rehan Jamil, Li Ming, Irfan Jamil, Rizwan Jamil

Application and Development Trend of Flue Gas Desulfurization (FGD) Process

A Review

GRIN Verlag

Bibliografische Information der Deutschen Nationalbibliothek:

Die Deutsche Bibliothek verzeichnet diese Publikation in der Deutschen National-
bibliografie; detaillierte bibliografische Daten sind im Internet über http://dnb.d-
nb.de/ abrufbar.

Imprint:

Copyright © 2013 GRIN Verlag GmbH
Druck und Bindung: Books on Demand GmbH, Norderstedt Germany
ISBN: 978-3-656-60139-5

This book at GRIN:

http://www.grin.com/en/e-book/269115/application-and-development-trend-of-
flue-gas-desulfurization-fgd-process

GRIN - Your knowledge has value

Der GRIN Verlag publiziert seit 1998 wissenschaftliche Arbeiten von Studenten, Hochschullehrern und anderen Akademikern als eBook und gedrucktes Buch. Die Verlagswebsite www.grin.com ist die ideale Plattform zur Veröffentlichung von Hausarbeiten, Abschlussarbeiten, wissenschaftlichen Aufsätzen, Dissertationen und Fachbüchern.

Visit us on the internet:

http://www.grin.com/

http://www.facebook.com/grincom

http://www.twitter.com/grin_com

International Journal of Innovation and Applied Studies
ISSN 2028-9324 Vol. 4 No. 2 Oct. 2013, pp. 286-297
© 2013 Innovative Space of Scientific Research Journals
http://www.issr-journals.org/ijias/

Application and Development Trend of Flue Gas Desulfurization (FGD) Process: A Review

Rehan Jamil[1], Li Ming[1], Irfan Jamil[2], and Rizwan Jamil[3]

[1]School of Physics & Electronic Information,
Yunnan Normal University,
Kunming, Yunnan, China

[2]College of Energy & Electrical Engineering,
Hohai University,
Nanjing, Jiangsu, China

[3]Heavy Mechanical Complex (HMC-3),
Taxila, Rawalpindi, Pakistan

ABSTRACT: In 1927, the limestone desulfurization process was first applied in the Barthes and Bansside Power Plants (total 120MW) beside the Thames River in UK to protect high-rise building in London. Up to now, over 10 desulfurization processes have been launched and applied. Based on the desulfurizing agent being used, there include calcium process (limestone/lime), ammonia process, magnesium process, sodium process, alkali alumina process, copper oxide/zinc process, active carbon process, ammonium dihydrogen phosphate process, etc. The calcium process is commercially available and widely used in the world, i.e. more than 90%. Flue gas desulfurization processes, survey made by the coal research institute under the International Energy Agency shows that the wet-process desulfurization accounts for 85% of total installed capacity of flue gas desulfurization units across the world. The wet-process desulfurization is mainly applied in countries, like Japan (98%), USA (92%), Germany (90%), etc. The limestone-gypsum wet desulfurization process, the most mature technology, the most applications, the most reliable operation in the world, may have rate of desulfurization of more than 90%. Currently, the flue gas desulfurization technology used at thermal power plants at home and abroad tends to be higher rate of desulfurization, bigger installed capacity, more advanced technology, lower investment, less land acquisition, lower operation cost, higher level of automation, more excellent reliability, etc. This paper briefs current situations and trends of flue gas desulfurization technology also append short descript of different type of FDG and their category.

KEYWORDS: Desulfurization Process, FGD technology, Wet-process limestone, lime-gypsum, thermal power.

1 INTRODUCTION

It was in UK that the 1st test on flue gas desulfurization in the world was carried out, following London Smog Episode (Dec., 1952). Up to the middle of 1960s, sulfur dioxide pollution had become a major concern related to environmental impact in UK, USA and former western Germany. Later, large-scale research was initiated in the USA and Germany [1]. Information shows that 20-plus means on flue gas desulfurization were among the USA's list of research in 1960s, which were applied in nearly 100 projects, consuming investment of billions of dollars [2]-[3]. Germany, the biggest coal-fired country in the Europe, ranks 1st on sulfur dioxide emission and leading to complaints from the neighboring countries. Thus, huge amount of human and material resources was made to address the problem. Japan, the 1st country in the world to apply the flue gas desulfurization unit in a large scale, had started operation of such unit as early as in the end of 1960s so that sulfur dioxide pollution in Japan was put into control since the middle and late of 1970s [1]. Applications of flue gas desulfurization unit in a large scale in USA, Germany and Japan demonstrate good yields at proper control. Despite of increasingly rising total installed capacity of power plants in the above 3 countries over the past few years, total amount of emission of sulfur dioxide drops on a year-on-year basis [3]-[4].

2 DEVELOPMENT COURSE OF FLUE GAS DESULFURIZATION TECHNOLOGY

2.1 INITIAL STAGE

Between the early of 1970s and end of 1970s, the 1_{st} generation of flue gas desulfurization technology based on the limestone wet process started to be applied in power plants, which included limestone wet process, lime wet process, MgO wet process, dual-alkali process, etc. The 1_{st} generation of flue gas desulfurization units was mainly installed in the USA and Japan [2].

2.2 DEVELOPING STAGE

Between the early of 1980s and end of 1980s, the flue gas desulfurization technology based on the dry process and semi-dry process was shaped, including spray drying, limestone injection into the furnace and activation of calcium process, circulating fluidized bed process, pipe jetting, etc. Amid the period, the wet limestone cleaning process experienced rapid development. Particularly, great changes had been made on use of single tower, as well as design and general arrangement of the tower.

Japan Ebara proposed the dry flue gas desulfurization process as early as in the early of 1970s and made joint research with Japan Atomic Energy Research Institute (JAERI). In 1980s, a pilot plant of 24,000 Nm^3/h in the Indiana of USA was built for testing. Currently, over 20 test units and pilot projects have been put in place in Japan, USA, France, Russia, etc. For the semi-dry flue gas desulfurization technology, it is outcome of joint efforts made by the American JOY and Denmark NIRO [2],[5]. Since the 1st industrial unit adopting such process was built in the North America in 1978, the process has been experiencing rapid growth, which is thus utilized in 12 countries and market shares have arrived at 10.29%. The rotary spray dry process (SDA), suitable for medium and low-sulfur coal in the past, is now upgraded to be compatible with high-sulfur coal. Since the process may function with good economic return, it is recognized as the 1980s'FGD technology [6].

2.3 MATURE STAGE

With entry into the 1990s, many developing countries (especially the Asian countries) have stipulated some emission standards designed to control acid rain. After development of two generations, FGD technology has advanced into a new period, i.e. simplification of redundant system in the desulfurization process greatly benefits operation reliability.

3 CURRENT SITUATION OF FLUE GAS DESULFURIZATION TECHNOLOGY

With FGD technology experiencing development and application over the past decades, some processes have been eliminated for technical or economic reasons. On the other hand, main processes, such as limestone/lime humidification process, flue gas circulating fluidized bed process, limestone injection into the furnace and activation of calcium process, spray drying process and upgraded humidification ash circulation NID process, etc., have grown up and become more mature, which may be demonstrated in the following areas:

3.1 HIGH DESULFURIZATION EFFICIENCY

The optimized wet process may yield desulfurization efficiency of more than 95%, spray drying and NID process 85%-90%, improved limestone injection into the furnace and activation of calcium process 85%, circulating fluidized bed process more than 90% in the case of same rate of utilization of absorbent as the wet process.

3.2 HIGH UTILIZATION

With in-depth understanding of chemical reaction principle of desulfurization process, appropriate control of reaction process and correct selection of structure material, as well as quality assurance from the desulfurization unit manufacturer, a high efficiency of utilization for the desulfurization system is achievable to ensure that it operates with the boiler in a synchronized step.

3.3 SIMPLIFIED PROCESS FLOW

Taking the limestone-gypsum wet process as an example, the process was first proposed by British Royal Chemical Corporation. Since 1970s, the process experienced 3 generations of development in the industrial applications over the past 3 decades, as a result of which the wet-process limestone/lime-gypsum flue gas desulfurization technology has well developed and widely applied [7]-[8].

The 1st generation of wet-process limestone/lime-gypsum flue gas desulfurization technology was put into the industrial applications since the early of 1970s. At that time, the desulfurization absorption unit was made up of 3 towers and 1 tank, i.e. Pre-scrubber, scrubber, oxidization tower and reaction interim storage tank [9], which is illustrated in fig.1

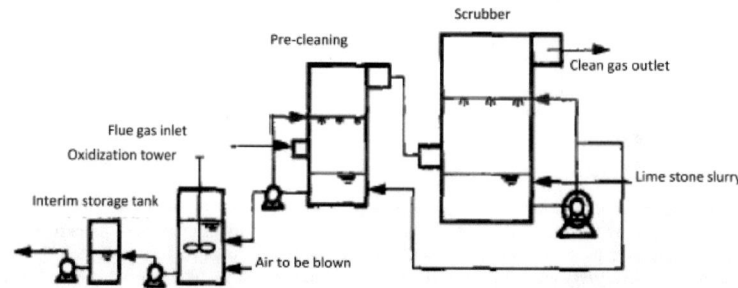

Fig. 1. *1st Generation of wet-Process Limestone/Lime-Gypsum Flue Gas Desulfurization System Schematic*

Approximately in the middle of 1980s, the 2nd generation of wet-process limestone/lime-gypsum flue gas desulfurization technology occurred. The absorption unit using the technology puts the pre-scrubber and scrubber together and eliminates the interim storage tank [10], as illustrated in fig.2

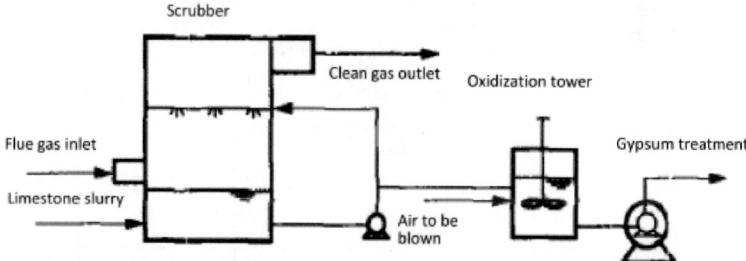

Fig. 2. *2nd Generation of wet-Process Limestone/Lime-Gypsum Flue Gas Desulfurization System Schematic*

In the 1990s, the 3rd generation of wet-process limestone/lime-gypsum flue gas desulfurization technology took shape, in which such unit was configured to integrate the pre-scrubber, scrubber and oxidization tower together to increase the velocity of flue gas, reduce the diameter of tower and acquire less floor space, as shown in fig.3. Utilization of the process may greatly cut the investment of desulfurization unit, approximately down 30-50% for initial investment; essentially address fouling, blocking problems, and enhance the system safety and reliability (≥95%); improve exposure of gas to liquid in the tower by making improvement on tower's internal parts to bring the rate of desulfurization to a higher level; make wider the commercial application of desulfurization byproduct based on research and development on recovery and utilization of such byproduct [11]-[12].

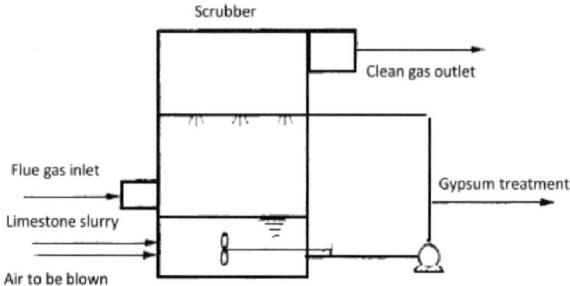

Fig. 3.   3rd Generation of wet-Process Limestone/Lime-Gypsum Flue Gas Desulfurization System Schematic

1）Lower system power consumption

2）Less investment and costs of operation

Over the recent years, simplified desulfurization process and optimized design parameters, system investment and operation costs are lowered by 1/3-1/2. Development and application of new wet desulfurization process such as integration of cooling tower with flue duct, deletion of GGH, multiple boiler with one tower, multiple tower with one boiler, etc. The flue gas desulfurization processes, survey made by the coal research institute under the International Energy Agency shows that the wet-process desulfurization accounts for 85% of total installed capacity of flue gas desulfurization units across the world. The wet-process desulfurization is mainly applied in countries, like Japan (98%), USA (92%), Germany (90%), etc. The limestone-gypsum wet desulfurization process, the most mature technology, the most applications, the most reliable operation in the world, may have rate of desulfurization of more than 90%. Its byproduct-gypsum may be reused, or disposed of [13]-[14].

4   DEVELOPMENT TRENDS OF FLUE GAS DESULFURIZATION TECHNOLOGY

With rapid progress in science and technology over the recent years, some new desulfurization technologies have been launched in foreign countries, but majority of which is still under test. Industrial applications in a large scale have not yet started [15]. Please see Table.1 for flue gas treatment technologies under R&D in foreign countries.

Table 1.   Current Situations on Research of Desulfurization Technology in Foreigner Countries

Desulfurization process	Category	Research institute	Desulfurization	Current situations
NO$_x$ SO technology	Dry process	US Department of Energy Pittsburgh Energy Technology Centre (PETC) and NO$_x$SO company	α-Al$_2$O$_3$ ball soaked with sodium carbonate	Industrial demonstration
SNO$_x$ technology	Dry process	Demark Haldor-Tops Φ A/S company	Ammonia gas	Industrial pilot test
DESONO$_x$ /REDOX process	Dry process	Germany Degussa A.G and Stadtwerk M ünster	Ammonia gas	Industrial pilot test
E-SOX process	Semi-dry Process	US Acurex company	Lime cream slurry	Industrial application in limited locations
Urea process	Wet process	Medeleev Institute of Chemical Technology of Russia	Urea	Industrial application in limited locations
LBL pHoSNO$_x$ process	Wet process	US Laurence Berkeley Lab and Bechel company	Sodium sulphite	Industrial pilot test
Seawater desulfurization	Wet process	US Marilyn Bechtel Group	Seawater plus lime	Industrial pilot test

There tends to a diversified, comprehensive and resource-oriented development of flue gas desulfurization process.

4.1 DIVERSIFICATION

There are over 200 flue gas desulfurization technologies in place. Recently, some new ones have taken shape, including calcium process, sodium process, magnesium process, manganese process, ammonia process, carbon process, seawater process, biological process, plasma process, membrane process, iron process, amine process, etc.

4.2 COMPREHENSIVENESS

Dust collection desulfurization integration technology, simultaneous desulfurization and denitration technology, multi-pollutant-aid removal technology are also on the agenda.

4.3 RESOURCE-ORIENTED

Now, various flue gas desulfurization technologies tend to turn byproduct into useful resources, such as recovery and utilization of gypsum in the limestone-gypsum desulfurization technology, recovery and utilization of sulfur ammonium, phosphorus ammonium in the ammonia desulfurization technology, recovery and utilization of sulfuric acid and sulfur in the activated carbon desulfurization technology, etc.

The general trend of wet-process limestone/lime-gypsum flue gas desulfurization technology, which is applied in the most areas, is to upgrade and optimize the system and make the equipment smaller so as to reduce investment and operation cost. Turning byproduct into useful resources and producing no secondary pollution are also the objective of the process.

5   CURRENT SITUATIONS AND DEVELOPMENT TRENDS OF FLUE GAS DESULFURIZATION TECHNOLOGY IN CHINA

China's research on flue gas desulfurization technology started at an earlier date, around 1950s. However, its development falls behind and is limited to purification of nonferrous metallurgical off gas and sulfuric acid tail gas.

5.1 DEVELOPMENT COURSE OF CHINA'S FLUE GAS DESULFURIZATION TECHNOLOGY

As early as in the 1950s, the nonferrous metallurgical industry started to make acid with flue gas with concentration of sulfur dioxide of higher than 0.5% through a purification process; carried out purification and recovery of sulfur dioxide in tail gas in sulfuric acid units; recovered and produced ammonium sulfate in nitrogen fertilizer plants by purifying sulfur dioxide tail gas; built and put into operation a number of large-scale plants [16].

Flue gas desulfurization test in coal-fired power plants started in the early of 1970s. Test and research on 6 different flue gas desulfurization processes were carried out in Shanghai Yangpu Power Plant, Shanghai Zhabei Power Plant, Sichuan Baima Power Plant and Douba Power Plant, Hunan 300 Plant and Songmuping Power Plant. Up to now, pilot test on spray drying process in the Baima Power Plant, phosphorous amine fertilizer process in the Douba Power Plant and iodine-containing activated carbon absorption flue gas desulfurization in the Songmuping Power Plant have been subject to technical evaluation. However, pilot test on the rest of 3 flue gas desulfurization technologies has not made progress for technical, management and economic reasons [17]-[18].

Currently, there are 8 kinds of flue gas desulfurization technology available in the country, including iodine-containing activated carbon process, sodium sulfite circulation process, amine-acid process, spray drying process. However, such processes are applied in some medium and small-size flue gas desulfurization units completing pilot test and are not advanced to a level suitable for large flue gas desulfurization units in coal-fired power plants.

To stimulate research and development of FGD technology across the country, the government scheduled to import a number of advanced technologies and units in a controlled way during the 7th and 8th Five-year Plan, as shown in Table.2 Such demonstration projects involve varieties of mature process, such as wet process from Japan, dry process and semi-dry process from the Europe and USA. Despite of advanced equipment, reliable operation and high level of automation, such projects cost too much on investment and operation, making difficult to be widely applied across the country. Deficiency of proprietary intellectual property rights may not benefit sustainable development of China's desulfurization technology [18].

Table 2. List of China Imported FGD Units

Imported by	Process	Amount of flue gas from boiler/(Nm^3.h^{-1})	Desulfurizing agent	Efficiency (%)	Operation since	Licensor
Shengli Oilfiled	Ammonia sulfur/ammonium process	2,100,000	NH$_3$ H$_2$SO$_4$	90	1979	Japan Toyo
Nanjing Steel and Iron Works	Basis aluminum sulfate process	51,800	Al$_2$(SO$_4$) Al(OH)$_3$	95	1981	Japan Dowa
Chongqing Luohuang Power Plant	Wet limestone-gypsum process	1,087,000	Limestone slurry	95	1992	Japan Mitsubishi Corporation
Shangdong Huangdao Power Plant	Simple spray drying process	300,000	Quicklime, Coal ash	70	1995	Power Development Co., Ltd of japan Inc.
Nanjing Xiaguan Power Plant	Limestone injection into furnace and activation of calcium process	795,000	Limestone	75	1997	Finland IVO
Taiyuan Generation Plant	Small high-velocity horizontal flow	600,000	Limestone	80	1996	Japan Hitachi
Guangxi Nanning Chemical Plant	Simple limestone-gypsum process	50,000	Ca(OH)$_2$	70	1996	Japan Kawasaki
Chengdu Thermal Power Plant	Electron beam process	300,000	NH$_3$	80	1997	Japan Ebara
Shangong Weifang Chemical Plant	Simple lime Gypsum process	100,000	Hydrate lime slurry	70	1995	Japan Mitsubishi Corporation

Over the recent decade, particular technological breakthrough was made on China's medium and small coal-fired industrial boiler flue gas desulfurization process, including direct injection of lime/limestone into the furnace and boiling bed limestone dry-process flue gas desulfurization, calcium alkali process, amine alkali process, sodium alkali process and magnesium alkali process, etc. Currently, there are over 40 coal-fired industrial boiler flue gas desulfurization processes, dozens of which operate well, with emission of sulfur dioxide up to the state permissible level. Please see Table.3 for China's own typical FGD technologies [18]-[19].

Table 3. Typical FGD Technologies in China

Desulfurization process	Developed by	Desulfurizing agent	Scale (Nm³/h)	Rate of desulfurization (%)	Current situations
Sodium sulphite process	Hubei No.300 Plant, etc.	Pure caustic	10,000	90	Complete pilot test
Phosphorous ammonia fertilizer process	Xi'an Thermal Power Institute, Sichuan University, etc.	Slag coke/phosphate ore	10,000	70-95	Industrial demonstration
Basic aluminium sulfate process	Chongqing Tianyuan Chemical Plant Power Station	Basic aluminium sulfate process	100,000	95	Industrial application
Managanses dioxide process	Sichuan University	Manganese dioxide slurry	7,000	>80	Complete pilot test and evaluation
Phosphate ore desulfurization process	Hunan University	Carbonate ore	5,000	63-70	Complete pilot test
Venturi desulfurization process	Environmental Protection Research Institute for Electric Power	Alkaline solution	75,000	63-70	Complete pilot test
Rotary spray drying process	Tsinghua University	Ca(OH)$_2$	70,000	85	Complete pilot test and not in industrial application
Limestone injection into the furnace and activation of calcium process	Shenyang Liming Company	Limestone	50,000	75	Complete industrial test
Desulfurization flue gas circulation sulfurization	Beijing College of Light Industry, Tsinghua University	Limestone	20,000	>80	Industrial demonstration
Activated coke process	Nanjing Electric Power Automation Equipment General Corporation	Activated carbon	200,000	95	Industrial demonstration
Iron Process	Dalian University of Technology, etc.	Iron	100,000	95	Industrial application
Circulating process	Chengdu West China Chemical Research Institute	Lon liquid	10,000	98	Complete industrial test

5.2 CURRENT SITUATIONS ON CHINA'S FLUE GAS DESULFURIZATION TECHNOLOGY

Recently, following China proprietary R&D and import, digestion, innovation, China's flue gas desulfurization industry had made considerable progress and the units adopting our proprietary process are sized to meet the target on reduced emission of sulfur dioxide set for the 11th Five-year Plan [25].

5.2.1 CONSIDERABLE PROGRESS MADE ON FLUE GAS DESULFURIZATION INDUSTRY IN THERMAL POWER PLANTS

By the end of 2008, the total installed capacity of flue gas desulfurization units in thermal power plants across the country has gone beyond 379 million kW, accounting for 66% of total installed capacity in thermal power plants. In 2008, China put into operation 100,000-level kW and higher desulfurization units of thermal power generating units, totaling 110 million kW, down by 5.2% compared with that in 2007 [20]. Currently, over 10 flue gas desulfurization processes have been applied, including limestone-gypsum wet process, flue gas circulating fluidized bed process, seawater desulfurization process, desulfurization dust collection integration process, semi-dry process, rotary spray drying process, limestone injection into the furnace and activation of calcium process, activated coke absorption process, electron beam process, etc. Like the situation in foreign countries, the limestone-gypsum wet-process flue gas desulfurization technology takes a leading position. Statistics show that such process has been utilized for more than 90% of thermal power plant projects already in operation, being constructed and signed [19]. China's flue gas desulfurization industry has grown up to a level able to complete design, manufacturing and general contracting of desulfurization projects sized for 100 million kW level [20],[23].

5.2.1.1 LOCALIZATION RATIO OF DESULFURIZATION EQUIPMENT UP TO MORE THAN 90 %

Key equipment in the limestone-gypsum wet-process flue gas desulfurization technology, like slurry circulating pump, vacuum belt filter, cyclone, boost-up fan, flue gas heat exchanger, flue gas damper, etc., are able to be designed and manufactured in the country. For example, a series of desulfurization slurry circulating pump manufactured by Shijiazhuang Pump Company have been applied in 96 desulfurization projects; desulfurization boost-up fan manufactured by Chengdu Power Machinery Factory has been applied in 100 desulfurization projects; gas-gas heat exchanger manufactured by Shanghai Boiler Plant has been applied in 60 desulfurization projects. In terms of purchase cost, equipment and materials adopting the limestone-gypsum wet-process flue gas desulfurization technology have localization ratio of around 90%, some flue gas desulfurization projects more than 95% and equipment for other processes more than 90%.

5.2.1.2 PROPRIETARY INTELLECTUAL PROPERTY RIGHTS AVAILABLE FOR MAIN FLUE GAS DESULFURIZATION TECHNOLOGY

Technology through Chinese own R&D and import, digestion and innovation, China has boasted main flue gas desulfurization technology of proprietary intellectual property rights suitable for 300MW-level thermal power generating unit, which has been tested in real units for more than 1 year. For example, China Power Investment Corporation Yuanda Environmental Protection Engineering Co., Ltd. Has had full picture of MHI double contact flow scrubber wet-process desulfurization technology, AEE spray tower wet-process desulfurization technology and AEE dry-process desulfurization technology, and then created YD-BSP wet-process flue gas desulfurization technology of proprietary intellectual property rights, which has been successfully applied in Nanyang Power Plant 2×300MW flue gas desulfurization project and Huaneng Haikou 2×125MW units flue gas desulfurization project; limestone-gypsum wet-process flue gas desulfurization technology of proprietary intellectual property rights developed by Suyuan Environmental Protection Engineering Co., Ltd. has been successfully applied in Taicang Harbor Environmental Protection Generating Co., Ltd. Phase-II 2×300MW flue gas desulfurization project; Beijing Guodian Longyuan Environmental Protection Engineering Co., Ltd. has boasted limestone-gypsum wet-process flue gas desulfurization technology of proprietary intellectual property rights following digestion, absorption and innovation of German technology, which has been successfully applied in Jiangyin Sulong Generating Co.,Ltd phase-III 2×330MW flue gas Desulfurization project [20]-[21].

After two years 'operation, the above three projects have been subject to the post-project evaluation. Experts view that flue gas desulfurization technology of proprietary intellectual property rights owned by three companies is technically mature, operationally reliable and highly compatible and thus is up to the internationally advanced level [22]. Even though we also own other processes of proprietary intellectual property rights, they are only suitable for thermal power units of 200MW and lower. Some units of this kind are just put into operation or being constructed, thus requiring test for a certain period.

5.2.1.3 AVAILABLE CAPABILITY OF GENERAL CONTRACTING OF FLUE GAS DESULFURIZATION PROJECT

By the end of 2008, about 50 enterprises, based on their technologies, funds and human resources, have experiences in undertaking turnkey contracting of flue gas desulfurization project of 100MW and higher, in which more than 20 ones get involved in total installed capacity of more than 3,000MW and 13 ones in that of 10,000MW. Beijing Guodian Longyuan Environmental Protection Engineering Co., Ltd., China Power Investment Corporation Yuanda Environmental Protection Engineering Co., Ltd, and other five companies total contractual capacity of more than 40,000MW, respectively [23].

5.2.1.4 CONSIDERABLE DECREASE OF COST OF DESULFURIZATION PROJECT

With considerable improvement of localization ratio of flue gas desulfurization equipment and market competition, cost of flue gas desulfurization project drops by big margin. For example, the cost of new thermal power generating unit flue gas desulfurization project of 300MW and higher has dropped from previous RMB 1,000 Yuan or above to existing RMB 200 Yuan or above per kilowatt. The cost of existing thermal power generating unit flue gas desulfurization project of 200MW and lower has dropped below RMB 250 Yuan per kilowatt.

5.2.2 KEY ENTERPRISES WITH RENOWNED BRAND EMERGED

For top 20 desulfurization companies in 2008, their contractual capacity accounts for 90.8% of the total, operation capacity for 78.3% of the total, operation capacity of the same year for 87.1% of the total. Compared with 2007, the ranking of top 10 desulfurization companies happen to no change, with Tsinghua Tongfang Environment Co., Ltd, and China Datang Technologies & Engineering Co., Ltd. among top 10. Please see Table.4 for overview of main desulfurization companies in China [24].

Table 4. Overview of Main Desulfurization Companies (by the end of 2008)

Item	Unit name	MW Total capacity by contract MW	Technical source	Proprietary technology
1	Beijing Guodian Longyuan Environmental Protection Engineering Co., Ltd.	68,829	German Steinmüller limestone-gypsum wet-process desulfurization technology	Longyuan wet-process flue gas desulfurization technology
2	China Boqi Environmental Protection Scientific and Technological (Holding) Co., Ltd.	52,496	Japanese Kawasaki spray tower technology	None
3	Wuhan Kaidi Electric Power Environmental Protection Co., Ltd.	49,700	Wet-process flue gas desulfurization technology from US B&W Co.	None
4	Fujian Longking Environmental Protection Co., Ltd.	42,360	Limestone-gypsum wet process, flue gas circulating fluidized bed dry process desulfurization technology from Germany LLB	None
5	China Power Investment Corporation Yuanda Environmental Protection Engineering Co., Ltd.	41,822	AEE wet-process spray and MHI double contact flow scrubber	BSP wet process
6	Zhejiang University Insigma Holding Electromechanical Engineering Co., Ltd	39,300	France ALSTOM limestone-gypsum wet process	None
7	Tsinghua Tongfang Environment Co., Ltd.	26,872	AEE wet-process spray double contact flow scrubber	None
8	Shandong Sanrong Environmental Protection Engineering Co., Ltd.	26,420	Wet process from Lurgi Bishchev	None
9	China Huadian Engineering Co., Ltd.	23,662	Wet-process flue gas desulfurization technology from US MET Co.	None
10	Zhejiang Tiandi Environmental Protection Engineering Co., Ltd.	19,050	Wet-process desulfurization technology from US B&W	None

5.3 DEVELOPMENT TRENDS OF FLUE GAS DESULFURIZATION TECHNOLOGY IN CHINA

Compared with foreign flue gas desulfurization technology, China's flue gas desulfurization technology is challenged by deficiency of mature proprietary flue gas desulfurization technology not applied in an economic and large scale in terms of boiler in some large coal-fired power stations; less factor given to recovery of sulfur resource; less application of large amount of byproduct, i.e., gypsum. In general, in view of existing problems, our flue gas desulfurization technology tends to:

5.3.1 ENHANCE CAPACITY FOR INDEPENDENT INNOVATION ON FLUE GAS DESULFURIZATION TECHNOLOGY

Up to now, only minority of desulfurization companies in the country boast flue gas desulfurization technology of proprietary intellectual property rights for 300MW and majority has to utilize the foreign technology for digestion and absorption but lack of capacity of re-innovation. On the other hand, we have to pay technology-introduced and used fees to foreign company. Preliminary calculations show that about RMB 320 million Yuan have been paid to foreign companies on licensing and RMB 300 million Yuan on royalty. Thus, improvement on our own independent innovation capacity of flue gas desulfurization technology is one of the trends guiding the development of flue gas desulfurization technology in the country [25].

5.3.2 REINFORCE DESULFURIZATION MARKET SUPERVISION

Recently, a great number of desulfurization environmental protection companies spring up like mashroom after rain to meet the requirements from dramatic expansion of desulfurization market. However, lack of supervision on market access, and detailed system in place to specify competence, talent, performance and financing for desulfurization companies lead to desulfurization companies of different quality mixed up together. Some flue gas desulfurization projects built by desulfurization companies are not acceptable. Furthermore, lack of or improper supervision on bid invitation or bidding process for flue gas desulfurization projects takes place. Some bid invitation or bidding processes are not put in place.

5.3.3 RAISE THE OPERATION RATIO OF DESULFURIZATION DEVICE

According to the insiders, the operation ratio of existing flue gas desulfurization units already built is less than 60%, thus not functioning as expected on reduced emission of sulfur dioxide. The reasons behind it are: first of all, some desulfurization companies highly rely upon foreign technology and equipment, as a result of which they are not able to have full picture of the technology being utilized, some systems are not designed well at the initial stage and individual equipment is difficult to be repaired in case of failure; secondly, desulfurization electric power price system for some old power plants is not put in place; thirdly, supervision or monitor on routine operation of desulfurization equipment is lacked due to slack law enforcement; fourthly, some power plants shutdown desulfurization systems for their own benefits.

5.3.4 RESOURCE-ORIENTED UTILIZATION

China's reserves on sulfur deposit ranks the 2_{nd} in the world. However, in 2004, our production of sulfur resource amounted to 8.1 million tons and imported 7.34 million tons; in 2006, they are 9.16 million tons and 9.5 million tons, respectively. China also boasts of abundant supply of gypsum. Gypsum, byproduct of desulfurization process, is not well applied, thus leading to secondary pollution. Therefore, the limestone-gypsum desulfurization process commercially available in foreign countries may not be suitable for the country. Instead, due factor shall be given to proper utilization of byproduct resource [26].

In addition, deletion of bypass flue duct, selection of gas discharging methods after desulfurization, deletion of GGH, economic operation and continuous monitor of flue gas are also key topics on flue gas desulfurization process under research in the country.

CONCLUSION

Flue gas desulfurization (FGD), is a desulfurization technology widely applied and highly efficient and is recognized by developed countries as the most economic and workable solution in the future. Flue gas desulfurization originated from the wet process experiment in 1930s. The desulfurization unit adopting limestone cleaning process in UK London Power Company and ammonia cleaning process in Canadian Cominco Company an earliest industrial were the earliest units of this kind. Flue gas desulfurization is designed to use desulfurizing agent to remove sulfur dioxide in flue gas based on the gas absorption, gas adsorption or catalytic conversion desulfurization mechanism. Amid the development and industrial application over the past few decades, over 200 desulfurization processes by making use of different desulfurizing agents or varying desulfurization mechanism had been made available in countries across the world, over 10 of which, however, are widely applied.

ACKNOWLEDGMENT

The authors would like to acknowledgement material support from China Power Investment Yuanda Environmental Protection Engineering Co., Ltd and financial support from Yunnan Normal University, Kunming, China.

REFERENCES

[1]   Beychok, Milton R, *Comparative economics of advanced regenerable flue gas desulfurization processes*, EPRI CS-1381, Electric Power Research Institute, March 1980.
[2]   Biondo, S.J. and Marten, J.C., "A History of Flue Gas Desulphurization Systems Since 1850," *Journal of the Air Pollution Control Association*, Vol. 27, No. 10, pp. 948–961, October 1977.
[3]   Beychok, Milton R., *Coping With SO₂*, Chemical Engineering/Desk book Issue, October 21, 1974.
[4]   Nolan, Paul S., "Flue Gas Desulfurization Technologies for Coal-Fired Power Plants," The Babcock & Wilcox Company, U.S., *the Coal-Tech 2000 International Conference*, November, 2000, Jakarta, Indonesia.
[5]   Rubin, E.S., Yeh, S., Hounsell, D.A., and Taylor, M.R., Experience curves for power plant emission control technologies, *International Journal of Energy Technology and Policy*, Vol. 2, No.3, pp. 223-233, 1/2, 2004.
[6]   Wu Zhong-biao, *Study on the wet and spray drying flue gas desulfurization*, Hangzhou: Zhejiang University, 1993.
[7]   Chen min-zhi, *Summary of Application of Fly Ash Desulfurization Technology*, Coal Ash China, 2002, 14(5):34-36.
[8]   J.D. Mobley, M.A. Cassidy and J. Dickerman, *Organic Acids can Enhance Wet Limestone Flue Gas Scrubbing*, Power Engineering, 90(5):32-35, 1986.
[9]   Kong Hua, Shi Zheng-lun, Gao Xiang, *Experimental Study of Wet Flue Gas Desulfurization in A Spray Scrubber*, Power Engineering, 21(5):1459-1463, 2001.
[10] Chang C S, Mobley J D., "Testing and commercialization of byproduct dibasic acid as buffer additives for limestone flue gas desulfurization systems," *Journal of the Air Pollution Control Association*, 33(10):955-962, 1983.
[11] C. T. Chi, J. H. Lester, *Utilization of adipic acid byproducts for energy recovery and enhancement of flue gas desulfurization Environmental progress*, 8(4):223-226, 1989.
[12] Wu Zhong-biao, Yu Shi-qing, Mo Jian-song, "Experimental Study of Desulfurization Process with Limestone Slurry Enhanced by Hexanedioic Acid," *Journal of Chemical Engineering of Chinese Universities*, Vol. 17 (5), pp. 540-544, 2003.
[13] Shi faen, Li zhen-tan, "Studies on Compound Additive of Wet FGD with Lime," *Sichuan Nonferrous Metal*, 18(3):27-29, 2003.
[14] Chen Zhijiang. *Theory and experiment research on the limestone flue gas desulfurization and additives in venturi scrubber*, Tianjing University press, 2000.
[15] Lawrence K. Wang, Norman C. Pereira, Yung-Tse Hung. Air pollution control engineering, *Handbook of Environmental Engineering volume1*, New Jersey: humana press Inc, 2004.
[16] Ru-shan Ren, Faen Shi, Yunnen Chen, Xue-min Huang, "Mass transfer model of the wet flue gas desulfurization with lime," *3rd International Environmental Pollution and Public Health Conference, June 14th to 16th, 2009 in Beijing China*.
[17] REN Ru-shan, HUANG Xue-min, SHI Fa-en, et al., "Research development on the Wet Flue Gas Desulfurization," *Industrial Safety and Environmental Protection*, Vol. 36 (6), pp. 14-15, 2010.
[18] Li Xianchun Han Renzhi, "Application and Forecast of Flue Gas Desulphurization Technology in Coal Fired Power *Plant*," *Angang Technology*, 2004-06.
[19] Feng yu, Yan li, *Economic Evaluation of Flue Gas Desulfurization Systems*; EPRI GS-7193; Electric Power Research Institute, February 1991.
[20] Feng Ling, Yang Jingling, Cai Shuzhong, *The Development of Flue Gas Desulphurization technology and its application*, Research Institute of Metallurgical Environmental Protection, Beijing, Environmental Engineering, 1997-02.
[21] Wang An, hang Yongkui, Chen Hua, Pu Yunxia, "Study on Micro Organism Method of Flue Gas Desulfurization," *International Journal of Chongqing Environmental Science*, Vol. 4, pp. 435-445, 2001-02.
[22] Pan Zhaoqun, "Review on Spray Dryer Absorption Flue Gas Desulfurization Technology," *International Journal of Chongqing Environmental Science*; Vol. 3, No. 8, pp. 145-153, 2003-08.
[23] Guo Ru-xin, "A Look at China's Potential for Flue Gas Desulfurization by Magnesium Oxide through Overseas R&D Progress," *Sulphur Phosphorus & Bulk Materials Handling Related Engineering*, 2009-02.

[24] Cao Gui-ping,Huang Bin,SUN Pei-shi,Cui Ya-wei, "Present Situation and Development Trend of SO_2 Flue Gas Technology in China," *Chinese Journal of Yunnan Environmental Science*, Vol. 9 No. 5, pp. 890-899, 2002-01.

[25] Su Qingqing, Yang Jiamo, "Development of flue gas desulfurization," *Chinese Journal of Wuhan Institute of Chemical Technology,* Vol. 543, No. 3, pp. 178-188, 2005-01.

[26] Zhang Yang-fan, Li Ding-long,Wang Jin, "The Development of Flue Gas Desulphurization Technology and its Application in China," *International Journal of Environmental Science and Management*, Vol. 5, No. 5, pp. 567-574, 2006-04.